RN, EFFICIENT
ION

)lan, BRE

This publication is sponsored by:

BRITISH READY-MIXED
CONCRETE ASSOCIATION

www.brmca.org.uk

The **Concrete** Centre

www.concretecentre.com

CONSTRUCT
concrete structures group

www.construct.org.uk

IHS | bre press

bre

CONTENTS

Inclined circular column formwork, PC6 Tower, Paris
(courtesy Outinord/G Hendoux)

Acknowledgements

Front cover photographs. Clockwise from left:
Guided climbing formwork for 19-storey residential block,
Stratford Eye, London (courtesy Doka);
Tunnel form structure – hotel in Lucaya, Bahamas (courtesy
Outinord/G Hendoux);
System column formwork (courtesy Peri);
Flying formwork (courtesy Peri).
Photo on title page. Slip form and jump form at Paddington
Central, London (courtesy PC Harrington)

BRE is committed to providing impartial and authoritative
information on all aspects of the built environment for
clients, designers, contractors, engineers, manufacturers
and owners. We make every effort to ensure the accuracy
and quality of information and guidance when it is
published. However, we can take no responsibility for the
subsequent use of this information, nor for any errors or
omissions it may contain.

BRE is the UK's leading centre of expertise on the built
environment, construction, energy use in buildings, fire
prevention and control, and risk management.

BRE, Garston, Watford WD25 9XX
Tel: 01923 664000
enquiries@bre.co.uk
www.bre.co.uk

BRE publications are available from
www.brepress.com
or
IHS BRE Press
Willoughby Road
Bracknell RG12 8FB
Tel: 01344 328038
Fax: 01344 328005
brepress@ihs.com

Requests to copy any part of this publication should be
made to the publisher:
IHS BRE Press
Garston, Watford WD25 9XX
Tel: 01923 664761
brepress@ihs.com

BR 495
© Copyright BRE 2007
First published 2007
ISBN 978-1-86081-975-9

INTRODUCTION

Concrete construction has gone through significant changes since the early 1990s and continues to develop. Innovations in formwork, concrete as a material, and reinforcement developments are just three of the contributors to what has become a significantly quicker, safer and less wasteful form of efficient construction.

A BRE study of innovation in concrete frame construction, funded by DTi, highlighted the huge impact that formwork innovation has had on speed and efficiency since the mid-1990s[1,2]. The study also highlighted how modern formwork systems for concrete structures have evolved to help meet the demands of the construction industry and its clients. These include demands such as cost, speed and efficiency, which have always been important to industry and will continue to be so. However, more recent drivers for sustainability have widened the demands to also embrace environmental and social concerns. Modern engineered formwork solutions have achieved step-change improvements in most of these areas compared with the bespoke timber-based types of formwork widely used up to the early 1990s.

This publication describes the main types of modern formwork systems that are widely available, considers their applications, advantages and main features related to health and safety, and sustainability performance. This information is intended to inform building industry professionals, particularly those involved with housing provision. Flats are forming an increasing percentage of new dwellings (currently about 45%) and many are constructed in concrete using efficient formwork systems.

Tunnel form at Allen Street Apartments, Dallas, Texas, USA (courtesy Outinord/G Hendoux)

Office development, Park Royal, London – lightweight aluminium formwork systems (courtesy Peri)

Formwork
A structure, usually temporary, used to contain poured concrete to mould it to the required dimensions and support until it is able to support itself. It consists primarily of the face contact material and the bearers that directly support the face contact material.

Falsework
Any temporary structure used to support a permanent structure until it is self-supporting.

AN OVERVIEW OF CONCRETE'S SUSTAINABILITY

Concrete is the largest commodity product in the world after water. However, its sustainability message has not been communicated as well as other materials despite the fact that concrete has many very sustainable features[3]:

- Modern concrete incorporates secondary materials which are by-products from other industrial processes, such as fly ash and ground granulated blastfurnace slag (GGBS), which minimize the usage of Portland cement. This is a good example of both economic and environmental drivers working together not against one another.

- The main bulk constituents of concrete (water and aggregates) are readily available close to most construction sites, keeping the impact of transporting raw materials low. Ready-mixed concrete plants aim to optimise production to use local materials most efficiently. There are approximately 1200 plants in the UK with an average delivery radius of less than 10 miles.

- Well designed and constructed concrete structures are very durable, potentially lasting hundreds of years. One often-cited example is the concrete dome roof of the Pantheon in Rome which is nearly 2000 years old. A more recent example, a terrace of 14 town houses at Marine Crescent, Folkestone, which has been redeveloped into 91 flats, highlights the adaptability of concrete structures. The original concrete building is over 150 years old and still providing modern housing[3].

- The ready-mixed concrete supply chain is decentralised and supports regional employment and development, an extremely important feature of sustainable communities.

- Concrete is 100% recyclable, as are the embedded steel reinforcing bars. Crushed concrete is routinely reused as fill in road construction instead of primary aggregates. Currently, little concrete waste goes to landfill as it has reuse value.

Completed jump form core and table form for a residential development, Manchester (courtesy Doka)

Marine Crescent, Folkestone, Kent, built in 1870 (courtesy The Concrete Centre)

BENEFITS OF CONCRETE FRAME BUILDINGS

While the primary focus of this document is on the construction process, it is useful to note the key performance benefits of concrete frame buildings.

Acoustics

The inherent mass of concrete attenuates airborne sound transmission. Additional finishes to walls and floors, required to meet Part E of the Building Regulations, are minimised or sometimes eliminated[4].

Adaptability

Concrete frame buildings can often be adapted to other uses. Holes can be cut through slabs and walls relatively simply, and methods for local strengthening are available if required. When a change of use is required this adaptability means that the structural frame is less likely to have to be demolished, extending its actual (as opposed to designed) life[5].

Mixed office/residential development, Tartu, Estonia using self-climbing jump form (courtesy Doka)

Office development, Park Royal, London – lightweight aluminium formwork systems (courtesy Peri)

Air tightness

Concrete frame buildings with flush column/slab details have surfaces against which seals can be made easily. Air tightness is critical to the current strategy for producing energy efficient buildings[5,6], by combining reduced air permeability of a structure with suitable ventilation. The concrete structure itself provides an effective barrier to air leakage.

Maintenance

A well designed and detailed concrete structure is almost maintenance free.

Performance in fire

Concrete is non-combustible and has a slow rate of heat transfer, which makes it a highly effective barrier to spread of fire. This inherent fire resistance means that concrete structures generally do not require additional fire protection. Normal levels of fire performance can be achieved through strength, continuity of reinforcement and good detailing of connections.

Concrete buildings and structures can often be repaired and returned to use after a fire[7-9].

Robustness

Reinforced concrete structures are normally considered to be inherently robust. This means that they are capable of withstanding explosions and other types of accidental loading. Disproportionate collapse does not occur when in-situ concrete frames are designed to the relevant UK codes of practice.

Security and vandal resistance

Concrete structures are strong, robust and well suited to providing a barrier to unauthorised or forced entry.

Thermal mass

A reinforced concrete structure has high thermal mass. This can be highly beneficial in moderating internal temperatures and, when used as part of a good overall design strategy, can lead to reduced operating costs and a comfortable environment for occupants[10-12].

Vibration control

The mass of concrete frames means that requirements for dynamic performance are readily satisfied, without the need for special provisions often associated with lightweight forms of construction[13,14].

School of Informatics, University of Edinburgh using lightweight panel system (courtesy Peri)

Mixed office/retail/leisure development at Bankside, Southwark, London using rail-guided climbing formwork for the core, with lightweight panel system for flat slabs (courtesy Peri)

CONCRETE ELEMENTS AND BUILDING SYSTEMS

In-situ concrete offers a wide range of construction solutions to designers, which can be chosen to suit a particular need[15]. The principle features of the main concrete frame elements used for building structures are noted below, except for fire resistance, robustness, thermal mass, cost effectiveness, speed of construction, sound control and durability of finish, which are generic to all concrete structures.

Flat slabs

Highly versatile elements widely used in construction, providing minimum depth, fast construction and allowing flexible column grids.
Markets: Residential and commercial buildings, hospitals, laboratories, hotels.
Perceived advantages: Design flexibility.

Ribbed and waffle slabs

Lighter and stiffer than equivalent flat slab solutions, offering enhanced advantages when slab vibration is an issue.
Markets: Vibration-critical projects, hospitals, laboratories.
Perceived advantages: Light weight which results in reduced foundation requirements. Reasonably shallow floors, excellent vibration characteristics, good service integration.

Beams and slabs

Commonly used for irregular grids and long spans where flat slabs are unsuitable. The beams can be wide and flat or narrow and deep.
Markets: Transfer structures, heavily loaded slabs, long spans.
Perceived advantages: Design flexibility.

Post-tensioned slabs

These are typically flat slabs, band beams or ribbed slabs, offering the thinnest slab type and allowing longer spans.
Markets: Commercial and residential buildings, hospitals, car parks, long spans.
Perceived advantages: Highly significant reductions in material use can be achieved.

Tunnel form structures

A form of cellular structure, in which vertical and horizontal members are cast at the same time. The term, which describes the formwork system, has become synonymous with the complete building shell.
Markets: Housing, hotels, hostels, student accommodation, prisons.
Perceived advantages: Dimensional accuracy of elements, safety, reduced formwork on site.

Queen Mary student village, University of London – tunnel form structure (courtesy Outinord/G Hendoux)

Concrete frame buildings normally contain both floor slabs and supporting structures such as walls or columns. A reinforced concrete 'core', such as a lift shaft or staircase, which has a primary structural function, may also be present. The formwork systems considered in this document are used to construct both the frame and the core.

The cores of most high-rise buildings are now constructed with modern efficient formwork techniques described in this publication irrespective of the structural materials used for the remainder of the construction.

TABLE FORM/FLYING FORM

Flying formwork (courtesy Peri)

Ground level assembly of table forms (courtesy Peri)

A table form/flying form is a large pre-assembled formwork and falsework unit, often forming a complete bay of suspended floor slab. It offers mobility and quick installation for construction projects with regular plan layouts or long repetitive structures, so is highly suitable for flat slab, and beam and slab layouts. It is routinely used for residential flats, hotels, hostels, offices and commercial buildings.

Generally, a series of individual falsework components including primary beams and props are connected to form a complete table, with plan area of up to about 100m². Trusses are often used, linked by cross-bracings. The formwork can be mounted on castor wheels or trolley units, allowing it to be moved horizontally. The basic construction sequence using this formwork is as follows:

- The assembled table formwork units are rolled in to position and sealed along the joints to form the floor to be cast.

- Steel reinforcement is fixed in place.

- Concrete is placed and cured.

- Once struck, the formwork units are lowered and rolled out from underneath the newly formed slab.

- They are then taken by crane and placed at the next position or level.

Process efficiency

- Speedy construction for large floor layouts.

- Fully assembled units can be manoeuvred quickly into place, rather than transporting individual components from one location to another and reassembling.

- High quality surface finishes can be achieved when appropriate quality control is used.

- The individual components of the formwork system are highly engineered and can be precisely adjusted.

- As the tables create large bay sizes the need for infill areas and decking joints is minimised. The high degree of repetition simplifies work practices.

- Reduced workforce requirement on site. However, the initial assembly of the formwork can be labour intensive depending on the size of the table unit.

- It is easier to plan construction activities due to the repetitive nature of the work.

Safety

- Table formwork systems can include standard health and safety features, such as guard rails. Edge protection is normally fixed after or during the assembly of the table form units.

- Table top decking and guard rails can be assembled at low levels and lifted on to the table falsework when complete. The falsework units can also be assembled at ground level, minimising work at height.

- Decking with non-slip surfaces can be used to enhance safety.

- Interconnected truss members provide a reasonably robust assembly and create a stable working platform.

- The repetitive nature of the work ensures that site operatives can quickly become familiar with health and safety aspects of their job.

Other sustainability features

- The formwork system is reusable with little waste generated compared to traditional formwork. The assembled units are intended for use throughout the duration of a project without dismantling.

- Increased speed and time efficiency on-site.

- Table form systems have been found to be very cost effective for structures with flat slabs.

- The repetitive nature of the work, combined with the engineered nature of the formwork, allows site teams to finely tune their operations, which in turn leads to minimal concrete wastage.

Other considerations

- The system requires enough space around the new construction to fly the table unit beyond the building line on every use.

- The supporting slab must be capable of carrying high loads at bearing locations; back propping may be needed underneath the slab.

- Lateral movement of tables must be carefully controlled using appropriate castors and trolleys.

- Adequate craneage is required for lifting the completed table unit. Lifting operations must be carefully controlled and comprehensive lifting plans are required.

- Planning is required to ensure sufficient space for assembling the table units.

- Assembly requires a workforce conversant with the system.

- Working platforms and safe access have to be provided separately.

- Some suppliers offer small table units that can be assembled off site and transported in. This can guarantee quality of workmanship and minimise delays due to poor weather. Smaller tables can offer greater flexibility and speed for movement and reuse, with less labour required. They can be lifted to the next level using C-hooks, whereas larger flying form units generally need crane slinging, which is a slower movement process, to gradually move units out of the building as slings are attached.

- Smaller tables often mean fewer sizes and simpler relocation and reuse.

- Pre-assembly can be particularly important on city centre sites with restricted assembly spaces.

- Table lifting systems and hoists are now available that allow smaller table units to be lifted between floor levels without the need for a crane. This frees up craneage for movement of reinforcement etc enabling greater productivity in poor weather.

Moving table forms using a C-hook, and proprietary column formwork incorporating access platforms (see pages 18–19) (courtesy Peri)

Secondary school, Solihull, Birmingham using table forms (courtesy Doka)

SYSTEM COLUMN FORMWORK

The column formwork systems now available are normally modular in nature and allow quick assembly and erection on-site while minimising labour and crane time. They are available in steel, aluminium and even cardboard (not reusable but recycled) and have a variety of internal face surfaces depending on the concrete finish required. Innovations have led to adjustable, reusable column forms which can be clamped on-site to give different column sizes.

The basic construction sequence using this type of formwork is as follows:

- The column forms are assembled and positioned over or enclosing the reinforcing bar cage.

- The forms are positively restrained and braced using props.

- Concrete is poured.

- Once the concrete has hardened sufficiently the formwork is stripped and moved to the next position manually or by crane. Disposable forms may be left in place for an extended period to aid curing and strength gain of the concrete before removal.

Process efficiency

- Metal column forms can be assembled and erected more easily than traditional formwork. Disposable forms come ready assembled to site.

- Increased speed and efficiency in construction.

- The simplicity of assembly and disassembly reduces the requirement for skilled labour.

- The highly engineered nature of the metal formwork systems enables precision adjustment to the formwork.

- High quality surface finishes are possible.

Raking column forms – Mercedes Benz Heritage and Technology Centre, Brooklands, Weybridge, Surrey (courtesy Peri)

Variable section column forms
(courtesy Outinord/G Hendoux)

Safety

- Metal formwork systems can have integral ready-to-use concreting platforms with guard rails and access equipment including ladders. This reduces the need for independent access.

- For systems with disposable formwork, working platforms for concreting have to be erected separately to allow safe access to the top of the column forms.

- Formwork systems are available which need to be worked only from one side. This could be an important safety consideration for columns situated at building edges and corners.

- Metal systems typically provide robust assemblies.

- The simplicity of the assembly process ensures that site operatives can quickly become familiar with health and safety aspects of their job.

- Normally these formwork systems require minimal use of power tools.

Other sustainability features

- The metal column forms are easy to clean and reuse with little waste generated compared to traditional formwork.

- The highly engineered nature of the metal formwork allows site teams better control of the operations, which in turn can lead to reduced concrete wastage.

- Disposable forms that are stripped and discarded after one use can often be recycled.

Other considerations

- The column forms are designed for specific maximum concrete pressures. The concrete placement rates have to be adjusted to keep the concrete pressure within the specified limits.

- The assembled formwork has to be restrained at the base properly to avoid displacement, and grout loss during concreting.

- In some metal systems, the push/pull props used to stabilise the column formwork are integral.

- Some systems can be moved on wheels rather than by crane.

- The formwork and access equipment can be moved in a single operation with some systems

- Some metal systems can be easily adjusted in plan size and height (by stacking additional panels on top of each other).

- Storage of disposable forms normally has to be carefully planned to avoid high temperature and long-term sun exposure of the forms. Care is needed in manual handling and crane operations to avoid damage to the forms from impact. These forms are not suitable for prolonged exposure to water or high moisture content.

- When stripping disposable forms, cutting tools have to be used with care to minimize damage to the finished concrete surface. Special forms that enable easy and quick stripping without the need for cutting are available.

System column formwork, shopping centre, Telford, Shropshire (courtesy Doka)

System column formwork (courtesy Peri)

HORIZONTAL PANEL SYSTEMS

Structural finishes and modular system formwork
(courtesy Peri)

Innovation in formwork has led to small lightweight modular systems that provide versatile formwork solutions on-site. Use of aluminium, high tensile steel, fibre glass, special plastics, etc. for different components has enabled lightweight units to be developed which can be safely manhandled to achieve a practical formwork system.

The components of the lightweight formwork systems are engineered to give robust and easy-to-handle solutions capable of dealing with both regular and irregular formwork areas. Minimising the number of different components in a formwork system allows mobility and quick installation of the formwork. The individual components often have multiple uses.

A range of lightweight modular formwork systems is available for the construction of most types of structure. They are suitable for any type of concrete frame construction.

Lightweight formwork systems used for slab construction generally consist of a series of interconnected falsework bays, independent props or system scaffolds and supporting pre-formed decking panels. These can include primary beams spanning between props and supporting a number of panels. The basic construction sequence using this type of formwork is as follows:

- The falsework and the decking panels are assembled on the ground or an existing slab to form the floor to be cast; joints in the decking are sealed.

- Reinforcement is placed.

- Concrete is placed and cured.

- Once struck, the formwork system is disassembled into individual components.

- Components are then moved manually and re-assembled at the next level or position.

Process efficiency

- The lightweight components enable larger areas of formwork to be assembled more easily than traditional formwork.

- Increased speed of construction.

- The light weight of the components and the highly engineered connection methods reduce the workforce requirement and increase the speed of assembly on site.

- The simplicity of assembly and disassembly of these modular systems reduces the requirement for skilled labour.

- Basic assembly is possible with minimal crane use as components can normally be manually handled to the next position or up to the next storey.

- The reduced weight of the formwork system means that the total load requiring back propping is reduced, indeed props within the system can be left in place. Pans (panels forming the underside of the slab) can be reassembled at the next position or on the floor above with spare drop-head props.

- The engineered individual components of the formwork systems enable precise adjustment to the formwork.

- Good quality surface finishes are possible when appropriate site controls are used. The numbers of joints mean that, where a very high quality of finish is required, special measures may have to be taken.

- These systems can be used on sites with space and access limitations.

Safety

- Erection can be carried out from below, largely eliminating working from height.

- These formwork systems have limited in-built standard health and safety features. Edge protection has to be fixed after or during the assembly when erecting or dismantling adjacent to slab edges.

- These systems provide a fairly robust assembly, which creates a stable working platform. However, for some systems additional stability may be required from the permanent structure.

- The simplicity of the assembly process and repetitive nature of the work in certain types of structure mean that site operatives can quickly become familiar with health and safety aspects of their job.

- These modular lightweight systems can be combined with wind-shield systems for slab-edge protection. Multiple floor height protection of the workforce can be provided to reduce the risk of falling from height. Slab edge screens can be either a steel mesh or fully enclosed by metal or ply sheeting, which provides better weather protection for the workforce.

Other sustainability features

- The formwork system is easy to clean and reuse with little waste generated compared to traditional formwork.

- The components can be moved quickly around the site. The individual system components such as props can be used for other purposes. This mobility and their modularity enable materials to be kept to a minimum on-site.

- Lightweight form systems can be very cost effective for certain types of structure.

- The engineered nature of the formwork (and the repetitive nature of work in certain types of structure) allows site teams better control of the operations, which in turn leads to reduced wastage.

Other considerations

- The support slab must be capable of carrying loads at bearing locations; back propping may be needed underneath the slab. In these instances propping can be carried out using the existing (drop-head) formwork components.

- Safe access has to be provided.

Modular system formwork (courtesy Peri)

VERTICAL PANEL SYSTEMS

Crane-lifted panel systems are commonly used on building sites to form vertical elements and usually consist of a steel frame with plywood, steel, plastic or composite facing material.

The systems are normally modular in nature, assembly times and labour costs are considerably lower than traditional formwork methods with far fewer components required. They offer greater opportunities for reuse for different applications on site.

Panel systems are extremely flexible and the larger crane-lifted versions can be used for constructing standard concrete walls, perimeter basement walls, columns and in conjunction with jump form climbing systems.

The basic construction sequence using this type of formwork is as follows:

- Panels are connected together flat on the ground to form larger individual crane-lifted units.

- After assembly of the units one face of the formwork is erected vertically in position and restrained/plumbed using props.

- Reinforcement is fixed against the erected formwork.

- The opposing face of the formwork is positioned and both form units are tied together.

- Concrete is poured.

- Once the concrete has hardened sufficiently the formwork is stripped and moved to the next position.

Process efficiency

- Assembly is very simple with panels connected and fixed using wedge clamps and hammers, reducing the requirement for skilled labour.

- Easily adaptable to varying structural geometries, wall heights, etc.

- Increased speed in construction, compared to traditional wooden formwork.

- The engineered nature of the panel formwork systems allows quick adjustment of the formwork.

Safety

- Working platforms, guard rails and ladders can be built into the completed units of formwork.

- The simplicity of the assembly process and repetitive nature of the work in certain types of structure ensure that site operatives can quickly become familiar with health and safety aspects of their job.

- Normally these formwork systems require minimal use of power tools.

Other sustainability features

- The forms are easy to clean and reuse repeatedly.

- Less waste is generated compared to traditional formwork.

Vertically stacked wallform 22.5m high – Disneyland, Paris (courtesy Outinord/G Hendoux)

Other considerations

- The panel systems are designed for specific maximum concrete pressures. The concrete placement rates have to be adjusted accordingly to keep the concrete pressure within the specified limits.

- Lightweight manhandled systems are available with steel or aluminium frames and plywood facing. These are commonly used in groundworks construction where a site crane is not always available.

- For a fair faced exposed concrete finish it may be necessary to line the panels with a secondary layer of material. This can still be more economical than traditional formwork, especially where the formwork needs to be adapted to varying site uses.

- For concrete where enhanced durability is required, controlled permeability formwork (normally in the form of a formwork liner) may be used. The formed concrete normally has a regular (textured) surface and an absence of blow holes and so can also be specified for architectural reasons.

- To simplify striking the formwork and moving it to the next location, stripping corners/panels have been developed. These are especially beneficial in confined lift shafts and stair cores where units can be lifted as four-sided boxes. With traditional formwork it may be necessary to break the four sides of the shaft into individual units, thus requiring more crane lifts and much higher labour costs in re-positioning/plumbing the formwork.

PRE-ASSEMBLED FORMWORK

- Concrete elements are sometimes specified to a high specification with a fair-faced or special finish. With this type of work and where unusual or complex structural geometries are encountered it is not always possible to use modular formwork systems.

- Various formwork suppliers are able to offer custom-built formwork pre-assembled in their factories.

- Where wall construction is repetitive and has many reuses it can be cost effective to use traditional formwork components that are pre-assembled rather than modular systems where the material costs can be higher.

- Pre-assembled formwork can reduce the requirement for space on-site.

System wall formwork (courtesy Peri)

JUMP FORM

Generally, jump form systems comprise the formwork and working platforms for cleaning/fixing of the formwork, steel fixing and concreting. The formwork supports itself on the concrete cast earlier so does not rely on support or access from other parts of the building or permanent works.

Jump form, here taken to include systems often described as climbing form, is suitable for construction of multi-storey vertical concrete elements in high-rise structures, such as shear walls, core walls, lift shafts, stair shafts and bridge pylons. These are constructed in a staged process. It is a highly productive system designed to increase speed and efficiency while minimising labour and crane time.

Systems are normally modular and can be joined to form long lengths to suit varying construction geometries. Three types of jump form are in general use:

- Normal jump/climbing form – units are individually lifted off the structure and relocated at the next construction level using a crane.

- Guided-climbing jump form – also uses a crane but offers greater safety and control during lifting as units remain anchored/guided by the structure.

- Self-climbing jump form – does not require a crane as it climbs on rails up the building by means of hydraulic jacks, or by jacking the platforms off internal recesses in the structure. It is possible to link the hydraulic jacks and lift multiple units in a single operation.

The basic construction sequence using this type of formwork is as follows:

- The formwork and the access platform are assembled on the ground.

- This combined formwork and access platform assembly is lifted using a crane and fixed to 'cast-in' anchors or tracks (climbing brackets) bolted to the wall elements below.

- Once the poured concrete is strong enough, the assembly is struck from the new/current concrete lift, and raised by crane to the next position. For self-climbing systems this is done using the hydraulic jacks.

- This whole cycle can be completed within a short time; the actual time depending on the size and complexity of the construction.

Guided climbing formwork for the core of Riverside Quarter residential development Wandsworth, London (courtesy Doka)

Process efficiency

- Fast construction can be achieved by careful planning of the construction process. Crane availability is critical for normal jump form.

- Self-climbing formwork cuts down the requirement for crane time considerably. By allowing the crane to be used for other construction work this may reduce the total number of cranes needed on site.

- The formwork is independently supported, so the shear walls and core walls can be completed ahead of the rest of the main building structure. This can help to provide stability to the main structure during its construction and can have the beneficial effect of taking the jump form core off the project critical path.

- High quality surface finishes are possible.

- Climbing forms can be designed to operate in high winds (when the use of a crane is less viable). This allows construction work to be carried out at reduced risk from adverse weather.

- The highly engineered nature of jump form systems allows quick and precise adjustment of the formwork in all planes.

- Some formwork systems can be used at an inclined angle, which is particularly useful on bridge pylons or where walls vary in thickness.

- A small but skilled workforce is required on site.

- It is easier to plan construction activities due to the repetitive nature of the work.

Safety

- Working platforms, guard rails, and ladders are built into the completed units of market-leading formwork systems. Complete wind-shield protection on platform edges is also possible.

- Self-climbing formwork systems are provided with integral free-fall braking devices.

- The completed formwork assembly is robust and provides a stable working platform.

- The reduced use of scaffolding and temporary work platforms results in less congestion on site.

- The setting rate of concrete in those parts of the structure supporting the form is critical in determining the rate at which construction can safely proceed.

- The repetitive nature of the work means that site operatives can quickly become familiar with health and safety aspects of their job. Formwork suppliers provide materials and resources to help train the labour force.

Other sustainability features

- The formwork system is easy to clean and reuse with little formwork waste generated compared to traditional formwork.

- Climbing formwork systems offer simplicity, safety and cost effectiveness for certain high-rise building structures.

- The repetitive nature of the work, combined with the engineered nature of the formwork, allows fine tuning of the construction operations, which in turn leads to minimal concrete wastage.

- Many repeated uses of formwork are possible before maintenance or replacement is needed, the number of uses depending on the quality of the surface finish of concrete specified.

Other considerations

- Jump form is typically used on buildings of five storeys or more; fully self-climbing systems are generally used on structures with more than 20 floor levels. However, a combination of crane-handled and self-climbing platforms can be viable on lower structures.

- Different systems have varying platform widths. Wider platforms allow the shutter to be fully retracted when struck for cleaning and steel fixing while narrower platforms use a tilting mechanism for shutter access.

- Trailing and suspended platforms are used for concrete finishing and retrieving cast-in anchor components (climbing brackets) from previous pours.

Climbing formwork system for the core of Royex House, Aldermanbury Square, London (courtesy Peri)

- Assembly and lifting operations for self-climbing formwork systems require personnel to be comprehensively trained to ensure competence. It is necessary to understand and comply with suppliers' method statements at all times.

- The raising operation must be carefully planned and coordinated, and access to the working area during lifting should be restricted to essential personnel.

- Self-climbing systems can be designed to incorporate additional platform levels above the main construction level. This enables steel fixing for the next casting cycle to start before striking and lifting the formwork, enabling multiple operations to be performed simultaneously and offering shorter construction cycle times.

SLIP FORM

Slip form is similar in nature and application to jump form, but the formwork is raised vertically in a continuous process. It is a method of vertically extruding a reinforced concrete section and is suitable for construction of core walls in high-rise structures – lift shafts, stair shafts, towers, etc. It is a self-contained formwork system and can require little crane time during construction.

This is a formwork system that can be used to form any regular shape of core. The formwork rises continuously, at a rate of about 300 mm per hour, supporting itself on the core and not relying on support or access from other parts of the building or permanent works.

Commonly, the formwork system has three platforms. The upper platform acts as a storage and distribution area while the middle platform, which is the main working platform, is at the top of the poured concrete level. The lower platform provides access for concrete finishing.

The basic construction sequence using this formwork is as follows:

- The formwork and the access platform are assembled on the ground.

- The assembly is raised using hydraulic jacks.

- As the formwork rises continuously, continuous concrete and rebar supply are needed until the operation is finished.

- At the end of the operation the formwork is removed using a crane.

- The entire process is thoroughly inspected and highly controlled.

Process efficiency

- Prudent and careful planning of construction can achieve high rates of production.

- The slip form does not require a crane to move upwards so the need for crane time is reduced. Concrete supply, on the other hand, can be heavily dependent on crane time or lift availability since volumes required are well below the capacity of normal concrete pumps.

- As this formwork operates independently, formation of the core in advance of the rest of the structure takes it off the critical path. This can help to provide stability to the main structure during its construction.

West India Quay, Canary Wharf, London – slip form core (courtesy PC Harrington)

- The availability of the different working platforms in the formwork system allows the exposed concrete at the bottom of the rising formwork to be finished, making it an integral part of the construction process.

- Certain formwork systems permit construction of tapered cores and towers.

- Slip form systems require a small but highly skilled workforce on site.

Safety

- Working platforms, guard rails, ladders and wind shields are normally built into the completed system.

- Reduced use of scaffolding and temporary work platforms results in a less congested construction site.

- The completed formwork assembly is robust and provides a stable working platform.

- The strength of the concrete in the wall below must be closely controlled to achieve stability during operation.

- The uniform and continuous nature of the work ensures that site operatives can quickly become familiar with health and safety aspects of their job. Formwork suppliers provide materials and resources to help train the labour force.

- High levels of planning and control mean that health and safety are normally addressed from the beginning of the work.

Other sustainability features

- The repetitive uniform nature of the work, combined with the engineered nature of the formwork, allows fine tuning of the construction operations, which in turn leads to reduced concrete wastage.

- The formwork system is reusable with little waste generated compared to traditional formwork.

- Slip form systems can offer safe and cost effective solutions for certain high-rise building structures.

Other considerations

- This formwork is more likely to be economical for buildings more than seven storeys high.

- Extensive planning and special detailing are needed as the process has little flexibility for change once continuous concreting has begun.

- Standby plant and equipment should be available though cold jointing may occasionally be necessary.

- The structure being slipformed should have significant dimensions in both major axes to ensure stability of the system.

- The setting rate of the concrete has to be constantly monitored to ensure that it is matched with the speed at which the forms are raised.

- Assembly and operations require personnel to be comprehensively trained to ensure competence. It is necessary to understand and comply with suppliers' method statements at all times.

Slip-formed 200m chimney at West Burton power station, Nottinghamshire (courtesy Bierrrum)

Slip form core for the O2 Arena, Millennium Dome, London (courtesy Byrne Bros)

TUNNEL FORM

Tunnel form is used to form repetitive cellular structures, and is widely recognised as a modern innovation that enables the construction of horizontal and vertical elements (walls and floors) together.

Significant productivity benefits have been achieved by using tunnel form to construct cellular buildings such as hotels, low- and high-rise housing, hostels, student accommodation, prison and barracks accommodation.

The normal dimensions of tunnel form units are 8 to 11 m long and 2.4 to 6.8 m wide. Individual units can be joined together to give tunnels of greater length.

The basic construction sequence is described in Ref 16. In summary, the forms are positioned, steel reinforcement is fixed and concrete poured and cured. When the concrete has hardened the forms are lowered and moved. The whole construction cycle can be as short as 24 hours.

Residential tunnel form structure
(courtesy Outinord/G Hendoux)

Tunnel form structure – hotel in Lucaya, Bahamas (courtesy Outinord/G Hendoux)

Process efficiency

- For the right type of building site and structure this is a very rapid form of construction.

- High quality surface finishes are possible.

- The engineered nature of the formwork can result in a high dimensional accuracy of the finished structure.

- Requires a smaller but multi-skilled workforce on site.

- It is easier to plan construction activities due to the repetitive nature of the work.

Safety

- Tunnel formwork systems include standard health and safety features[17] such as guard rails. Market leading systems often have edge protection built-in.

- Most tunnel forms are delivered to site partly assembled, resulting in less manual handling. Assembly is completed at ground level.

- The completed formwork assembly is robust and provides a stable working platform.

- The repetitive nature of the work ensures that site operatives can quickly become familiar with health and safety aspects of their job. Formwork suppliers often provide materials and resources to help train the workforce.

- The need to use power tools for assembly is moderate.

Other sustainability features

- The formwork system is easy to clean and reuse with little waste generated compared to traditional formwork.

- Tunnel form systems can be very cost effective for certain types of structure.

- The repetitive nature of the work, combined with the engineered formwork, allows site teams to finely tune their operations, which leads to minimal concrete wastage.

Other considerations

- Tunnel formwork is particularly economical for projects of 100 or more cellular units, but there are examples exist abroad where tunnel forms are used for low-rise housing and much smaller cellular frame construction.

- Sufficient space is required to allow safe removal of the tunnel from the structure being constructed.

- Planning is required to ensure adequate space on site for transportation, storage, assembly and disassembly.

- Assembly requires a workforce fully conversant with the system.

Student accommodation at University of East Anglia – tunnel form structure (courtesy The Concrete Centre)

SUMMARY

Formwork systems used for concrete frame construction have continued to develop significantly since the early 1990s. The major innovations have focused on on-site efficiency of production, health and safety, and environmental issues, driving the concrete construction industry towards ever increasing efficiency.

Different formwork systems provide a wide range of concrete construction solutions that can be chosen to suit the needs of a particular development.

Traditional formwork for concrete construction normally consisted of bespoke solutions requiring skilled craftsmen. This type of formwork often had poor safety features and gave slow rates of construction on-site and huge levels of waste – inefficient and unsustainable.

Modern formwork systems, which are mostly modular, are designed for speed and efficiency. They are engineered to provide increased accuracy and minimize waste in construction and most have enhanced heath and safety features built-in.

Slipformed core of HQ3, Canary Wharf, London (courtesy PC Harrington)

The main systems in use are table form/flying form, system column formwork, horizontal and vertical panel systems, jump form, slip form and tunnel form. This guide sets out their key features – process efficiency, safety, sustainability and other considerations – in order to help construction professionals to take advantage of them to achieve modern, efficient concrete construction.

Wall formwork, RAF Museum, Cosford, Staffordshire (courtesy A-Plant Acrow)

ACKNOWLEDGEMENTS

Thanks are due to the following for their assistance in preparing this guide.

A-Plant Acrow
102 Dalton Avenue, Birchwood Park, Warrington, Cheshire WA3 6YE
Tel: 01925 281000
E-mail: enquiries@aplant.com
www.aplant.com

Doka UK Formwork Technologies Ltd
Monchelsea Farm, Heath Road, Boughton Monchelsea, Maidstone, Kent ME17 4JD
Tel: 01622 749 050
E-mail: uk@doka.com
www.doka.com

EFCO UK Ltd
22-28 Meadow Close, Ise Valley Ind Estate, Wellingborough, Northamptonshire NN8 4BH
Tel: 01933 276775
E-mail: uk@efcoforms.com
www.efcoforms.com

Ischebeck Titan Ltd
John Dean House, Wellington Road, Burton-on-Trent, Staffs DE14 2TG
Tel: 01283 515677
E-mail: sales@ischebeck-titan.co.uk
www.ischebeck-titan.co.uk

Outinord International Ltd
105-6 New Bond Street, London W15 1DN
Tel: 0203 214 2055
E-mail: andrew.sims@outinord.net
www.outinord.net

PERI Ltd
Market Harborough Road, Clifton upon Dunsmore, Rugby CV23 0AN
Tel: 01788 861600
E-mail: info@peri.ltd.uk
www.peri.ltd.uk

RMD Kwikform
Brickyard Road, Aldridge, West Midlands WS9 8BW
Tel: 01922 743743
E-mail: info@rmdkwikform.com
www.rmdkwikform.com

SGB Formwork
Harsco House, Regent Park, 299 Kingston Road, Leatherhead RH17 6HL
Tel: 01372 38 1300
E-mail: info@sgb.co.uk
www.sgbformwork.co.uk

Special thanks to Bierrum International Ltd, Byrne Bros (Formwork) Ltd and PC Harrington (Slipform International) for photographs of slip form systems.

REFERENCES

1 Nolan, É. Innovation in concrete frame construction 1995–2015. IHS BRE Press, Garston. BR 483. 44pp.

2 Nolan, É. Innovation in concrete frame construction, IHS BRE Press, Garston, August 2005. BRE Information Paper 11/05. 6pp.

3 Case study on Marine Crescent, Folkestone is at www.concretecentre.com/publications. More information is available at *www.cementindustry.co.uk* and *www.sustainableconcrete.org.uk*.

4 BRE/CIRIA. Sound control for homes. IHS BRE Press, Garston, 1993. BRE Report BR 238/CIRIA Report 127. 129pp

5 Dimitriloupolou C et al. Ventilation, air tightness and indoor air quality in new homes. IHS BRE Press, 2005. BR 477. 64pp.

6 BSRIA. Air tightness specifications. BSRIA, Bracknell, 1988. BSRIA Specification 10/98. 8pp.

7 Lennon T, Rupasinghe R, Waleed N, Canisius G and Matthews S. Concrete structures in fire: performance, design and analysis. IHS BRE Press, Garston. 2007. BR 490. 80pp.

8 Lennon T. Structural fire engineering design: materials behaviour – concrete. IHS BRE Press, Garston, 2004. BRE Digest 487. 8pp.

9 The Concrete Society. Assessment and repair of fire-damaged concrete structures. Camberley, 1990. Concrete Society Technical Report 33.

10 Harrison H W et al. Designing quality buildings – a BRE guide. IHS BRE Press, Garston, 2007. BR 487. 360pp.

11 Barnard N, Concannon P and Jaunzens D. Modelling the performance of thermal mass. IHS BRE Press, Garston, 2001. BRE Information Paper 6/01. 12pp.

12 De Saulles T. Utilisation of thermal mass in non-residential buildings. The Concrete Centre, Camberley, 2006. CCIP-020. 90pp.

13 Arup. Hospital floor vibration study, comparison of possible floor structures with respect to NHS vibration criteria. Research report, 2004, available from *www.concretecentre.com/publications*.

14 Willford M R and Young P. A design guide for footfall induced vibration of structures. The Concrete Centre, Camberley, 2006. CCIP-016. 82pp.

15 Industrial brochures and case studies are available from The Concrete Centre (*www.concretecentre.com/publications*).

16 The Concrete Centre. High performance buildings – using tunnel form concrete construction. The Concrete Centre, Camberley, 2004. TCC04/02. 8pp.

17 CONSTRUCT. Safe erection and dismantling of falsework and formwork. CONSTRUCT Concrete Structures Group, Camberley, 2007. (*www.construct.org.uk/hands.asp*).

(Courtesy A-Plant Acrow)

Back cover photographs. Special column formwork – station at Roissy II airport, France (courtesy Outinord/G Hendoux); Rail climbing/screening system on an 18-storey residential tower, Newham, London (courtesy Peri)